ARCTIC ANIMALS

FUN THINGS TO LOOK FOR IN THIS BOOK

LIFT-A-FACTS

Every page has two flaps that can be lifted to reveal fun facts about the animals. One flap has a fascinating fact about a physical characteristic of the animal, and the other flap tests your knowledge about the animal with a Do You Know? question.

SIZE COMPARISONS

Look for the size icon on each page. The featured animal is compared to a 4-foot tall child.

GLOSSARY WORDS

You can learn more about the meaning of certain words by looking in the glossary located in the back of the book.

Polar Bear

It's cold in the North Pole but polar bears like living there! Their thick fur and layers of fat keep them warm as they float on ice and swim in the frigid ocean. They use their large, slightly webbed front paws to paddle through water.

Unlike most bears, polar bears do not **hibernate,** but mother polar bears will dig dens in the snow to keep their cubs safe and warm.

Polar bears' **fur camouflages** them in snow, but under their fur, they have **black skin** that **soaks** up the sun to help keep them **warm**.

DO YOU KNOW...

how polar bears keep from **slipping** on the **ice**?

Polar bears have **no natural enemies** in the wild. This means **no** other animals **kill** polar bears for **food.**

DO YOU KNOW...

how **fast** puffins can **fly**?

The puffin's beak is about one inch long.

Inches 0 1 2 3 4 5 6 7 8

Centimeters 0 1 2 3 4 5 6 7 8 9 10 11 12 13 14 15 16 17 18 19 20 21 22

Atlantic Puffin

Puffins' **feathers** contain **special oils** that make them **WATERPROOF.**

This bird might look like a penguin, but it's not—it's a puffin! The puffin has a colorful beak that fades in the winter months and brightens during the spring. Puffins spend most of their time in the water. If they are not swimming, they are resting by floating on the waves.

A baby puffin is called a **puffling.**

Male and female puffins **rub** their beaks together to make a **"puffin kiss."**

Harp Seal

Harp seals don't play musical instruments—they get their name from the harp-shaped spots on their fur. Harp seals spend most of their time in the cold ocean hunting for fish and crustaceans. They can stay underwater up to fifteen minutes without coming up for a breath.

DO YOU KNOW...

why harp seals have **nostrils** that **close**?

Baby harp seals are born with **white fur** to help **camouflage** them in the snow. As they get older, their fur **turns** a **grayish color.**

Harp seals can **easily** dive down **600 feet.** That is like diving off a **tall city building!**

The **eyes** of harp seals have **lenses** that help them **focus** on objects that are **far away**.

Wolverines' teeth and jaws are **so strong** they often eat even the **bones** and **teeth** of their prey.

Wolverines do not travel in **groups.** They are solitary animals—that means they travel and live **alone.**

The wolverine's tail can grow up to 14 inches long.

Inches 0   1   2   3   4   5   6   7   8

Centimeters 2 3 4 5 6 7 8 9 10 11 12 13 14 15 16 17 18 19 20 21 22

Wolverine

Beware of the wolverine! This powerful animal may look like a small bear, but it is not related to the bear—it is the largest member of the weasel family. Even though wolverines will eat fruit and plants during the summer months, they are far from vegetarians, preferring to eat rabbits, skunks, porcupines, and other rodents.

Mother wolverines make **dens** or other **shelters** in the ground where they give **birth** to their babies, usually **two** or **three** at a time.

11 12 13 14 15

26 27 28 29 30 31 32 33 34 35 36 37 38

Arctic Fox

Arctic foxes live in the tundra, freezing cold areas without any trees. They can survive in some of the coldest temperatures in the Arctic—as low as 50 degrees below zero! How do these foxes survive such cold temperatures? They have one of the warmest coats of fur in the animal world. Their short ears, legs, and muzzle help them hold in heat so they don't get cold.

Depending on the season, there isn't always **much food** in the **tundra** for an Arctic fox to eat, so sometimes they eat a **polar bear's** **"leftovers."**

Arctic foxes have **white** fur in the **winter** and **brown** or **gray** fur during the **summer** months.

DO YOU KNOW...

why Arctic foxes have **fur** on the **bottom** of their paws?

Both **mother** and **father** Arctic foxes **raise** their **young.**

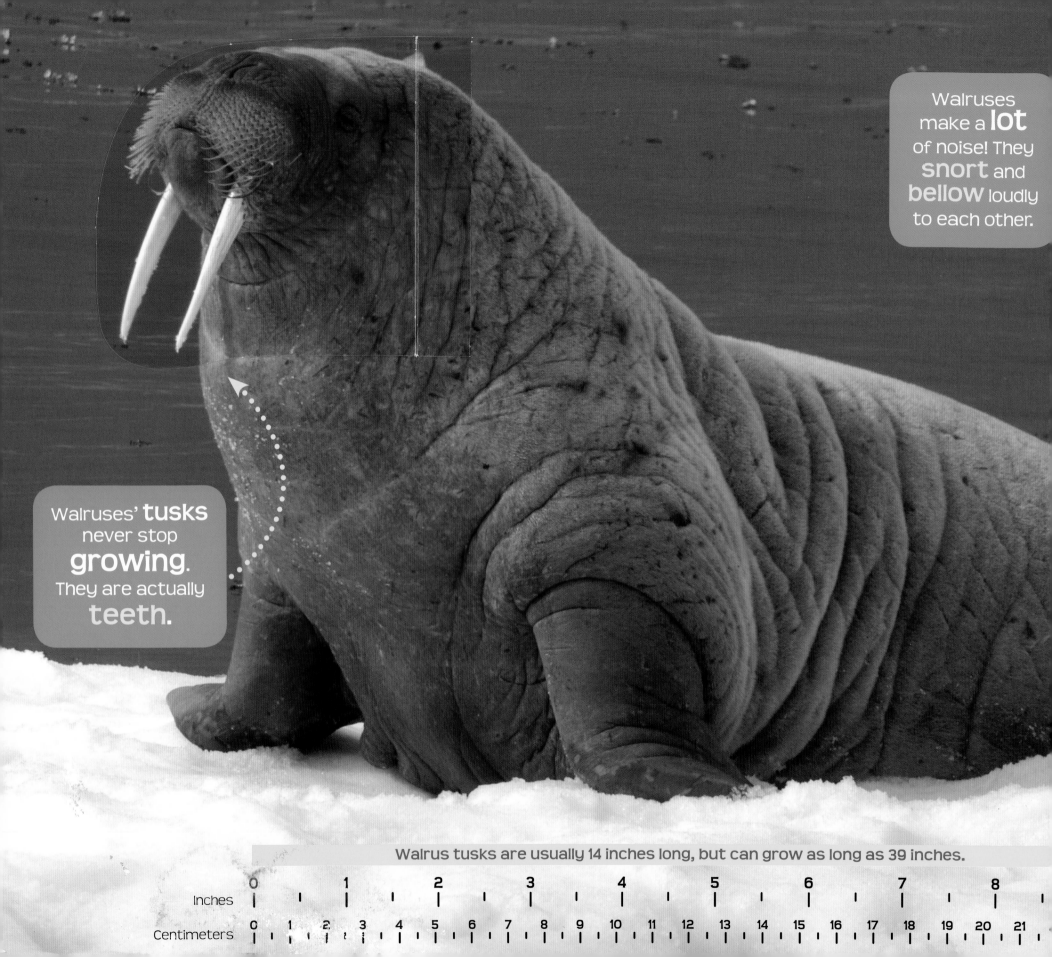

Walruses make a **lot** of noise! They **snort** and **bellow** loudly to each other.

Walruses' **tusks** never stop **growing**. They are actually **teeth**.

Walrus tusks are usually 14 inches long, but can grow as long as 39 inches.

Inches   0   1   2   3   4   5   6   7   8

Centimeters   0 1 2 3 4 5 6 7 8 9 10 11 12 13 14 15 16 17 18 19 20 21

Walrus

It's hard to mistake a walrus for another animal when you see its long tusks and whiskers! Walruses are mammals that can be found near the Arctic Circle. They have a lot of blubber to keep them warm in the cold ocean and they have the ability to slow their heart rate so they can stay underwater for nearly 10 minutes without coming up for air!

DO YOU KNOW...

what makes the skin of walruses **change color**, varying from dark brown to a subtle pink?

Mother walruses keep their **babies safe** from crowds of walruses by putting them up on **rocks** and **ice floes**.

10 11 12 13 **14** 15

26 27 28 29 30 31 32 33 34 35 36 37 38

Beluga

It is not hard to see why beluga whales are also known as white whales. Their white coloring makes them one of the most recognizable whales in the sea. Compared to some other whales, belugas are on the small side. Beluga whales are found mostly in the Arctic Ocean and they live and travel in small groups called pods.

Belugas **breathe** air through **BLOWHOLES,** but they can stay **underwater** for up to **20** minutes.

Beluga whales are **not** born **white.** It takes up to *nine years* before a beluga is **pure white.**

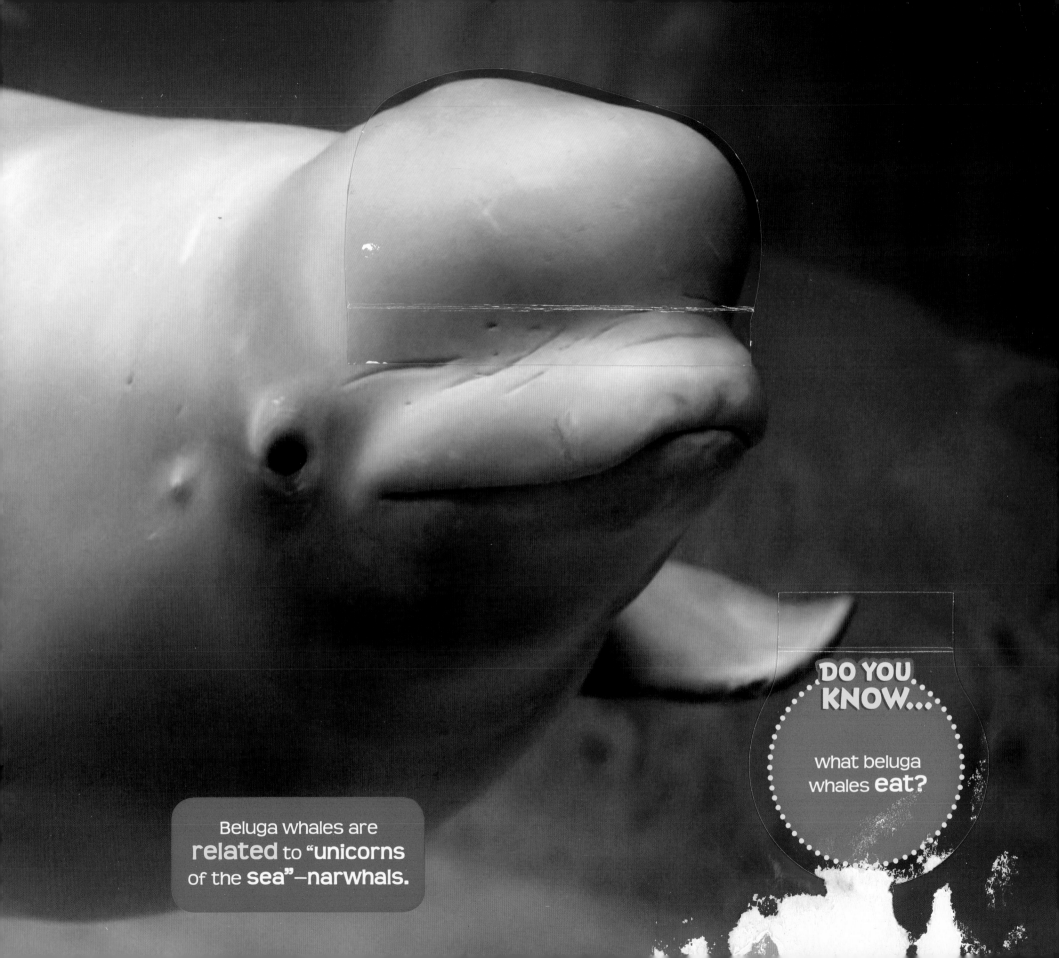

DO YOU KNOW...

what beluga whales **eat?**

Beluga whales are **related** to "**unicorns** of the **sea**"—**narwhals.**

DO YOU
KNOW...

what a
female moose
is called?

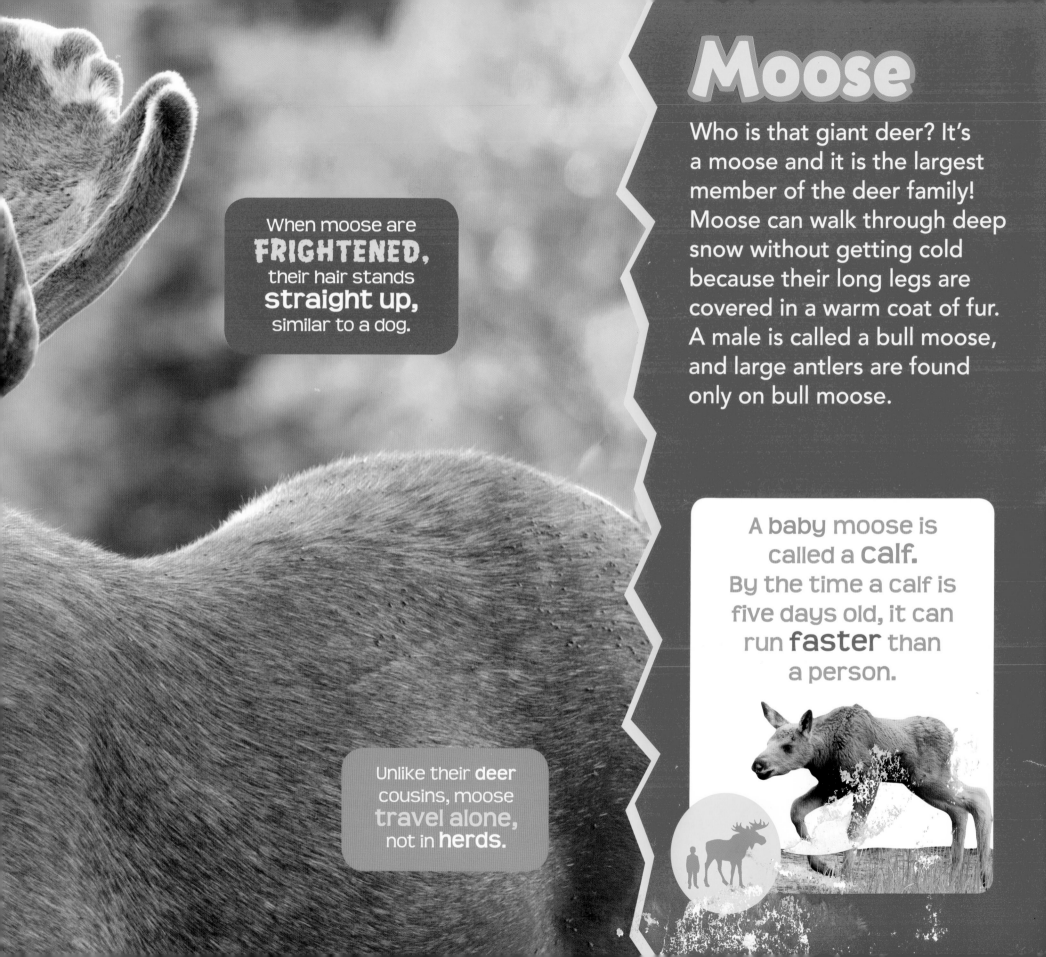

Moose

Who is that giant deer? It's a moose and it is the largest member of the deer family! Moose can walk through deep snow without getting cold because their long legs are covered in a warm coat of fur. A male is called a bull moose, and large antlers are found only on bull moose.

When moose are **FRIGHTENED,** their hair stands **straight up,** similar to a dog.

A baby moose is called a **calf.** By the time a calf is five days old, it can run **faster** than a person.

Unlike their **deer** cousins, moose **travel alone,** not in **herds.**

Snowshoe Hare

Is this a cute white rabbit? No, it's a snowshoe hare! Hares and rabbits are related to each other but they are not exactly the same. Most hares have longer ears and bigger feet than rabbits. Newborn baby hares are born with fur and can see, while baby rabbits are born hairless and blind. Snowshoe hares are shy mammals that can be found in cold climates.

Unlike rabbits, snowshoe hares **rarely** go underground in burrows. Instead, they hide in piles of leaves or hollow logs.

DO YOU KNOW... how snowshoe hares **communicate** with each other?

Snowshoe hares' feet can be up to 5.5 inches long.

Inches 0 1 2 3 4

Centimeters 0 1 2 3 4 5 6 7 8 9 10 11

Snowshoe hares are **fast.** They can **move** at speeds up to **30 miles** per hour!

Snowshoe hares are **white** in the **winter** and **brown** in the **summer.**

Glossary

Blubber: A layer of fat on sea mammals that helps keep them warm.

Burrow: A hole or tunnel dug in the ground by an animal as a home or for hiding.

Camouflage: An animal's coloring that allows it to blend in with its natural surroundings.

Climate: The long-term weather conditions in a specific area.

Crustacean: A creature with a hard outer shell and many legs, that often lives in water, like shrimp, crabs, and lobster.

Den: An area underground used by an animal for rest or hibernation.

Hibernate: When an animal goes underground or into a nest or tree hollow to "sleep" for the winter months. Animals that hibernate are able to slow their heart rate and breathing.

Mammal: A warm-blooded animal that has fur or hair.

Muzzle: The mouth and nose that projects from an animal's face.

Prey: An animal that is killed by another animal for food.

Rodent: A type of animal that includes mice, rats, and squirrels.

Webbed: Toes that are connected by thin skin called membrane.

Scientific Consultant

Jennifer Gresham
Director of Education
Zoo New England

Photo Credits

Dreamstime
iStockphoto
Shutterstock